"Food Is Power"

A Community Leaders Guide To The Global Food Revolution

Food is power, and whoever has the food has the power. – Dr. Don Matison

By:
Dr. Daniel Daves, Ph.D.

"Food Is Power" A Community Leaders Guide To The Global Food Revolution

ISBN 978-0-9895919-4-2

Printed in the United States of America
Bible translations used include: NIV, KJV, GW & MSG

Dr. Daniel Daves: Masters of Missiology, D. Min., Ph.D. Christian Administration, Logos Bible College & Graduate School, Jacksonville, FL USA
www.doctordanieldaves.com

To order more copies for you or your group, go to www.mighteagle.com or contact Mighty Eagle Publishing at info@mightyeagle.com.

www.globalfoodproviders.com

Table of Contents

Introduction:

Food is power, and whoever has the food has the power. – Dr. Don Matison

With passion and fervency I implore you to take to heart the contents of this booklet. It's the supplemental book to "Global Food Revolution" and it's directly focused towards Christian community leaders of all kind: Pastors, business people, NPO leaders, government officials, educators tribal chiefs and leaders of faith.

We stand at the precipice of a cataclysmic global food crisis that will affect you and those under your influence, if it's not already. Our world is changing quickly. Populations are exploding, weather patterns are permanently shifting and top world health leaders are shouting -

"Take immediate evasive action or suffer the consequences!".

Quote: *"The numbers tell a deeply worrying story. It is a race against time - we need to act now to save the lives and livelihoods of the millions at the brink of starvation today. "* -

Ertharin Cousin, Executive Director United Nations World Food Programme

Global useable farmlands are being depleted. Farmers are retiring. Global populations are sky rocketing. And while we need to double our global food production, we are actually in decline. This spells disaster for mankind, especially decimating the poor, the widow and the orphan. Those who are "at risk" have no capacity move out of the way, to make needed local adjustments, start mass agricultural projects or receive critical education to head off the coming food collapse.

Quote: "*The world needs to produce at least 50% more food to feed 9 billion people by 2050. But climate change could cut crop yields by more than 25%. The land, biodiversity, oceans, forests, and other forms of natural capital are being depleted at unprecedented rates. Unless we change how we grow our food and manage our natural capital, food security—especially for the world's poorest—will be at risk.* – United Nations Food Report - World Bank, 2016

Time is short and strong global and community leadership is needed in order to head off the greatest crisis of our lifetimes.

My friend. We must not be late to this current grand world event. A billion or more lives are counting on us. The miraculous will unfold if we will arrive on time. But we must not miss the season. We must seize the moment now.

Esther 4:14 (NIV) *And who knows but that you have come to royal position for such a time as this?"*

Indeed, we have come to the kingdom for such a time as this.

- Dr. Daniel Daves

Get our global food revolution informational updates at www.globalfoodproviders.com.

Chapter 1: Time Is Short. Food Insecurity Is Here.

If there was ever a grand opportunity for leaders to seize hold of global food security, it's right right now. We are currently losing 3,900 people per day from starvation around the world. Those numbers are set to double, then double again, and again in the coming years. All the statistics, scientific data, and global food organizations are shouting loudly:

"Get ready for the biggest global food disruptions of our life times."

Global food security is so weak that the slightest disruption in food supply will empty grocery store shelves, strip warehouse supplies and send the masses into panic. Current weather disruptions are stripping farms of their farm yields, bankrupting entire farm regions, and forcing agriculture to move to new growing zones. Science says that the weather disruptions are only the prelude to the coming "Grand Events" which could destroy entire national and regional food supplies.

What can be done to stop this? As a matter of fact, your leadership is required to prepare, train and focus your community towards "food security". While you may not have been required to understand or learn food security in the past, undoubtedly your family, business, church and community need your leadership in this field NOW! To put this off will be to miss your opportunity to secure your community.

The reason that I wrote this supplemental booklet to Christian leaders is to warn you of pending global disaster, and to help you get in front of this problem before it overtakes you and your community. Food security and sustainability are absolutely possible if we can see the need and develop a local plan of action. But we need to do it now rather than later. Agricultural food projects, training and proper learning curves can take years to successfully develop. If we wait until the storm is upon us, we will be like the five foolish virgins who didn't have oil in their lamps.

Matthew 25:6-12 (GW) *6 "At midnight someone shouted, 'The groom is here! Come to meet him!' 7 Then all the bridesmaids woke up and got their lamps ready. 8 "The foolish ones said to the wise ones, 'Give us some of your oil. Our lamps are going out.' 9 "But the wise bridesmaids replied, 'We can't do that. There won't be enough for both of us. Go! Find* *someone to sell you some oil.'* [10] *"While they were buying oil, the groom arrived. The bridesmaids who were ready went with him into the wedding hall, and the door was shut.* [11] *"Later the other bridesmaids arrived and said, 'Sir, sir, open the door for us!'* [12] *"But he answered them, 'I don't even know who you are!'*

The Holy Bible tells the story of how God's people can potentially miss the biggest events of our lifetimes. I shudder in fear at the thought that we could potentially miss this one. This problem is going to require a new mindset. It will take on a new form of action and leadership. It will require a new bottle.

Luke 5:37-39 (NIV) [37] *And no one pours new wine into old wineskins. If he does, the new wine will burst the skins, the wine will run out and the wineskins will be ruined.* [38] *No, new wine must be poured into new wineskins.* [39] *And no one after drinking old wine wants the new, for he says, 'The old is better.'"*

This story makes it clear that God cannot fill an old, antiquated system with a new idea. We are continually reminded by the Word of God to renew our minds, goals, and vision in order to stay in position for a fresh move of the Holy

Spirit. We are also reminded that a person who has drank from the old blessings of the past will not necessarily desire the new because they naturally desire to live in the glories of the past.

There is no time for living in the past or for embracing old dynamics, ideologies, or paradigms. A new day is at hand. God is looking for a new bottle that can contain a fresh move of His Holy Spirit that will touch humanity on a grand global scale. This new global food security crisis is a great church opportunity that will take us far outside the walls of our local church buildings. This crisis and corresponding opportunities will thrust us into the harvest fields of industry, agriculture, food preparation and distribution. We will find ourselves deeply involved in humanitarian, industrial, political, educational and business segments of society where billions of people live and work twenty four hours per day, seven days per week. The gospel message will be delivered in new bottles, because this new day has come upon us while the world is living in utter darkness.

Isaiah 60:1-4 (KJV) *[1] Arise, shine; for thy light is come, and the glory of the LORD is risen upon thee. [2] For, behold, the darkness shall cover the earth, and gross darkness the people: but the*

LORD shall arise upon thee, and his glory shall be seen upon thee. [3] And the Gentiles shall come to thy light, and kings to the brightness of thy rising. [4] Lift up thine eyes round about, and see: all they gather themselves together, they come to thee: thy sons shall come from far, and thy daughters shall be nursed at thy side.

The greatest darkness becomes the greatest opportunity for those of whom the glory of the Lord rises upon. We don't want to miss this opportunity that's standing in front of us. Seize the moment. Seize the day. The glory of the Lord is rising upon you, and the world will see His light clearly as you lead the way to salvation for your community.

Matthew 11:12 KJV *[12] And from the days of John the Baptist until now the kingdom of heaven suffereth violence, and the violent take it by force.*

Seize the moment, and seize the day.

Chapter 2: Be Sober And Vigilant

The facts are certain that great changes are coming to planet earth. All the facts and statistics are posted clearly in the book “Global Food Revolution”. And while we attempt to gain insight to the future and how the grand food security crisis will affect our families, businesses, communities and the church at large, let us as leaders, be sober and vigilant in our leadership.

1 Peter 5:8-11 KJV *[8] Be sober, be vigilant; because your adversary the devil, as a roaring lion, walketh about, seeking whom he may devour: [9] Whom resist stedfast in the faith, knowing that the same afflictions are accomplished in your brethren that are in the world. [10] But the God of all grace, who hath called us unto his eternal glory by Christ Jesus, after that ye have suffered a while, make you perfect, stablish, strengthen, settle you . [11] To him be glory and dominion for ever and ever. Amen.*

Sober: *Temperate in any way; not extreme or extravagant, serious, solemn, grave, or sedate, not exaggerated or distorted: the sober truth,* *characterized by reason, sanity, or self-control; showing mental and emotional balance.*

Vigilant: *Staying watchful and alert to danger or trouble.*

We are nearing the 2,000 year anniversary of the birth of the church of Jesus Christ. Church leaders are gathering together "en masse" to make sure that the gospel has been preached to every nation, tribe, tongue, race and people group before that anniversary date arrives around 2030. We are nearing the finish line and must utilize every opportunity that presents itself in this coming decade.

Hosea 6:1-2 KJV *[1] Come, and let us return unto the Lord : for he hath torn, and he will heal us; he hath smitten, and he will bind us up. [2] After two days will he revive us: in the third day he will raise us up, and we shall live in his sight.*

We are living in the season of spiritual and natural paradigm shift, where the 2nd day is swallowed up by the third day *(a day as a thousand years, 2 Peter 3:8)* and where the plowman overtakes the reaper and the harvesters will overtake those who are sowing seed.

Amos 9:13-15 (KJV) *[13] Behold, the days come, saith the LORD, that the plowman shall overtake the reaper, and the treader of grapes him that soweth seed; and the mountains shall drop sweet wine, and all the hills shall melt. [14] And I will bring again the captivity of my people of Israel, and they shall build the waste cities, and inhabit them; and they shall plant vineyards, and drink the wine thereof; they shall also make gardens, and eat the fruit of them. [15] And I will plant them upon their land, and they shall no more be pulled up out of their land which I have given them, saith the LORD thy God.*

Indeed an acceleration of the kingdom is upon us where many things begin to speed up. What used to take a decade will now take a year. Business models that demanded five years now will be completed in 30 days. Those who are sowing will be bumped from behind by the harvesters because of the acceleration of seedtime and harvest. Are you prepared for the new wine paradigm shift? Do you have a new bottle or wineskin in place for this new day?

The global community needs your leadership during this new paradigm in the earth. God is calling you to step up by faith into a glorious new calling that will reach far beyond your wildest dreams. The gospel is going to be preached in places never heard before, and it will be received by hungry people who are praying right now for your arrival and leadership. Even the most hardened of nations, the most remote locations and the most resistant of people will be calling for you to come.

Seize the moment. Seize the day.

Chapter 3: Seize The Day With Action

My friend, I want to strongly encourage you to make new considerations for your life, family, church, business, neighborhood, nation and the global community. Food insecurity is coming at society like a speeding freight train. We will all be affected one way or the other in the near future. Whether by rising prices of food, disappearance of some foods in your area, or seeing the affects of starvation and homeless first hand, the train is set to arrive at full speed in a community near you. It will have no mercy upon those it hits. How do we know?

Science, mathematics, and factual evidence shows population explosion and usable farm lands diminishing at a rapid pace. All world health and food organizations including the United Nations are crying out for your leadership and action.

Experts are calling for a mass infusion of business investment capital to help create the 50% more food supplies needed in the coming years. They are asking for entrepreneurial intervention in the agricultural and scientific worlds. They are calling for mass action around the world in order to shield those most "at risk" from present and future trauma, hunger and mass starvation. This is something that your leadership can define, identify and activate among your relational capacity. There are entrepreneurs, inventors, developers, business people, and youth who are full of energy and are looking for a wave to catch, an opportunity to seize, and a global initiative to be a part of. They are all around us. It's part of your calling and leadership capacity to encourage and thrust these people into the harvest fields of the new day.

Look around you and you will find agricultural experts, farmers, home gardeners and those with the proverbial "green thumb". You can find those with a knack for gardening and farming, and those whose lives are clearly headed towards food production, processing and distribution. Begin to empower them to seize the day and build programs that will help feed the poor, the widow, and the orphan while helping to create more nutrition and calories for the masses in their circle of influence. They will pay for their programs by selling food and nutrition that they are producing. There will be plenty left over for the poor. Help these people get educated and trained, if needed, and get them into position to accelerate during this new season of crisis and gospel opportunity.

I want you to consider using my book, "Global Food Revolution" as a viable resource to help arm and train your people. This book is full of all the statistics, technologies, directives, and ideas to empower them to search for local answers and to find the correct paths for your community. The answer will be different for every locality because of altitude, environment, seasonal cycles, weather pattern shifts, governmental oversight, etc.. But the answer for your community is available if you will pursue it. Let's pursue it together.

I believe that the Holy Spirit will show His leaders exactly what the first and next steps are. He will cause great faith to come as He unveils your local plan for food security coupled along with the gospel message. Sometimes the directives may not seem naturally like the right thing to do. But remember Isaac who planted his seed in the midst of a great famine and the Lord blessed him. He was right on time with the change of season as he planted in faith.

Genesis 26:12-14 (NIV) *12 Isaac planted crops in that land and the same year reaped a hundredfold, because the LORD blessed him. 13*

The man became rich, and his wealth continued to grow until he became very wealthy. [14] *He had so many flocks and herds and servants that the Philistines envied him.*

We all need to be making multi-tiered, multi-leveled plans to educate and prepare our base to be able to provide some level of sustainability regarding food growing, preparation, processing, packaging, and distribution. As you seek the Lord, you will receive a wisdomatic plan of action that you can implement among your people, family, and community. If you are a leader involved in multiple cities or nations, your scope and purpose must reach out much further. But the plan can be accomplished in record time and to the benefit of all.

Chapter 4: Wind In Your Sails

There is strong indication that the Holy Spirit is saying to His church, "I am going to put the wind in your sails.". But we need to hoist the sails into position – NOW! And for him who has no ship, it's time to buy or build one.

If you were to build a sailing vessel and then wait for the wind to come fill the sails, without that wind you will not move, or as they say, you are "dead in the water". This is a principle of business and timing that I have learned over decades.

I have built my fair share of ships, businesses, ideas and projects in my youth that didn't move like I had hoped. I wasn't in time with the idea or the wind currents. But I have learned through the past to listen to the Spirit of God, to put my proverbial finger into the wind, and catch its movement and direction before building anything.

John 3:8 (NIV) [8] *The wind blows wherever it pleases. You hear its sound, but you cannot tell where it comes from or where it is going. So it is with everyone born of the Spirit."*

The wind is filling the sails of "Global Food

Security" now, and you can build a vessel and ride it to great success in this season. What am I talking about? Food security for your people, neighborhood, community and your circle of influence. I am talking about business ventures focused in this direction that will have the full attention of global populations. This is the time for Ideas and concepts created for this great day of need. This is the time for people to arise to the occasion and seize the moment.

Now more than ever, I am running across people that are looking to buy farms, grow food, or learn how to sustain their family with a garden. These are people who've never farmed before, never gardened, or been involved in food. Many of them say they sense that God wants them to get involved with food production. Why? He is positioning His people for coming food shortages. If you listen carefully, you will hear the same conversations among your people. Why? God is putting wind in the sails of this ship called Global Food Security.

Chapter 5: Food Is Power And Influence

When I was a young youth pastor in church located in a small Kansas town, I learned a great deal from my pastor as we lived and worked in the wheat fields. He would always say things like:

"Food is power, and whoever has the food has the power."

He would also hold up a wheat seed and say:

"How many fields of wheat are in this one seed? None, if you don't plant it in the ground."

Through these "church in the wheat field" moments, I was learning more than I understood at the time. I would soon learn that food really IS massive power and influence. I would see how I could bring in a semi truckload of food to our community to give away, and I could immediately get 20 - 30 churches to work together with me. With food I could open government doors and walk into school systems that are normally closed to the gospel message.

Food really is power! And whoever has it, has great power and authority to unify people and help transform a community. I have found that the gospel easily rides in on the wings of food distribution to the poor, needy, the widow and the orphan.

In the Holy Bible, Jesus spoke to Peter and made a statement three times to him. I believe that anything Jesus needs to say three times must be very important.

John 21:15-18 (MSG) *[15] After breakfast, Jesus said to Simon Peter, "Simon, son of John, do you love me more than these?" "Yes, Master, you know I love you." Jesus said, "Feed my lambs." [16] He then asked a second time, "Simon, son of John, do you love me?" "Yes, Master, you know I love you." Jesus said, "Shepherd my sheep." [17] Then he said it a third time: "Simon, son of John, do you love me?" Peter was upset that he asked for the third time, "Do you love me?" so he answered, "Master, you know everything there is to know. You've got to know that I love you." Jesus said, "Feed my sheep. [18] I'm telling you the very truth now: When you were young you dressed yourself and went wherever you wished, but when you get old you'll have to stretch out your hands while someone else dresses you and takes you where you don't want to go."*

There are great secrets unveiled in this Biblical text. Jesus was giving Peter some final instructions before He would be crucified. Three times he told him to feed the lambs and sheep. I am convinced that this was a command to provide not only spiritual feeding but also natural food. Jesus taught the disciples through His ministry how to turn a few loaves and fish into a "Heaven blessed" divine supply for multitudes.

Somehow, there is a connection between hungry people, bread, fish, the gospel message and a faith filled believer who is willing to act.

I have found through decades of feeding the poor that a great door of supernatural provision opens when we open up our natural supplies of food to the poor. While I don't understand fully how it happens, I know that it absolutely works, because it's a law of the kingdom of God. When we give to the poor using natural food, supernatural windows of Heaven open up over us and those receiving the food.

I can tell you countless stories of supernatural multiplication of food that has happened frequently in our ministry of giving. I remember one time when in Philippines when we were preparing to feed around 200 people at a little village church rally we were doing for a local pastor. We had paid for the food in advance and the ladies were preparing it in the kitchen. Suddenly the local pastor came to us with fear in his eyes and told us that around 600 people were at the rally and needed to eat. I asked how much it would cost to provide for the rest of the people and he told me $150. I pulled out the cash to simply take care of it. But when they went to get the supplies, the local store was out of food. There was nothing available for purchase to feed these extra 400 people. All we could do was pray. And then we miraculously fed all 600 people that day, with bountiful portions of food on each plate, leaving enough for all of the workers to take huge portions home for their families. It was a modern day miracle, and the servers saw the hand of God fill the pots with more food every time they dipped the serving spoons in. It was a wonderful blessing for all, attributed to the goodness of God! We had many salvations and miracles that day which all started with the miracle of food multiplication. And the pastors faith exploded from that point forward.

I remember in the beginning of our feeding ministry, when we were launching out by faith to reach a high crime neighborhood in St. Louis,

Missouri in the USA. Previously we had always struggled to keep our little church pantry stocked for the poor. But the Lord had weighed heavily on my heart to believe Him for a semi truckload of food to be given away. I remember standing in front of the church and telling them that I was going to believe the Lord for a semi truckload of food to reach Fairgrounds Park. This was the 2nd highest crime area in our city. Some young people had just been gunned down in that park. The community was deeply fearful. We were going to roll in there and break the back of fear, terror and poverty by preaching the gospel. But we needed a truckload of food to bring the people out into the open so we could preach the message of hope.

I didn't know how much a truckload of food would cost. I didn't know how much it weighed or where it could come from. But I knew that I needed a truckload and I locked on to it by faith. I started making phone calls, asking questions, and pushing the envelope. Within a week I had a network of 26 churches ready to help preach the gospel and distribute the food. I made friends with a wonderful Bishop who allocated all of his local resources to make the event happen just one block from his main church at Fairgrounds Park. And suddenly I had a semi truckload of food donated to us from a food bank half way across the country. We set everything up, and by the time the event unfolded, I had three semi truckloads of food donated to our event. One of them was full of frozen high Dollar meats. That day we preached the gospel to over 800 people, we played music, had a mobile health Clinic, police/crime education, free haircuts, clowns and children's activities. We gave away huge bags of nice, free clothing, had specific prayer and medical booths, and we gave over 100,000 pounds of food away that day. The poor were hiring taxi cabs to take 150 - 200 pounds of food home with them. Our church volunteers were packing church vans and automobiles with food and clothes to help people get home with the bounty. It was a miraculous event and the start of our "truckload faith" of giving food away. Never again would we struggle to fill the church pantry. From that point forward, it was truckloads and pallets full of nutrition. Thank you Jesus!

When you start feeding the poor, you gain immediate influence with them, with politicians, business owners, community leaders, media professionals and anyone who has a heart of compassion for the underdog. It's really supernatural how "food is power" and "food is influence". I believe that this is the secret that

Jesus was trying to get across to Peter when He asked three times, "Do you love me? Feed my sheep." He was showing Peter the supernatural way to unlock the kingdom of Heaven for divine supply, and to unlock the hearts of people of influence towards the gospel.

Jesus fed the 5,000 (Matt. 14:15-20) and the 4,000 (Matt. 15:32-38) through miraculous supply. He told the disciples, "You feed them." He was activating their faith to see more than they had in their hands. He wanted them to take natural food, and through giving it to others it would supernaturally increase and multiply. It worked for Him. It worked for them. And it will work for you. Do you want to see the supernatural in your life, family, business, or ministry? Start giving food to the poor and asking the Lord to bless and multiply it. Watch out! You might be in line for semi truckloads, plane loads, train loads, and shiploads of miraculous food supplies.

Jesus feeds the 4,000: Matthew 15:32-38 (MSG) *32 But Jesus wasn't finished with them. He called his disciples and said, "I hurt for these people. For three days now they've been with me, and now they have nothing to eat. I can't send them away without a meal—they'd* *probably collapse on the road."* [33] *His disciples said, "But where in this deserted place are you going to dig up enough food for a meal?"* [34] *Jesus asked, "How much bread do you have?" "Seven loaves," they said, "plus a few fish."* [35] *At that, Jesus directed the people to sit down.* [36] *He took the seven loaves and the fish. After giving thanks, he divided it up and gave it to the people.* [37] *Everyone ate. They had all they wanted. It took seven large baskets to collect the leftovers.* [38] Over four thousand people ate their fill at that meal.

Jesus feeds the 5,000: ***Matthew 14:15-21 (NIV)*** [15] *As evening approached, the disciples came to him and said, "This is a remote place, and it's already getting late. Send the crowds away, so they can go to the villages and buy themselves some food."* [16] *Jesus replied, "They do not need to go away. You give them something to eat."* [17] *"We have here only five loaves of bread and two fish," they answered.* [18] *"Bring them here to me," he said.* [19] *And he directed the people to sit down on the grass. Taking the five loaves and the two fish and looking up to heaven, he gave thanks and broke the loaves. Then he gave them to the disciples, and the disciples gave them to the people.* [20] *They all ate and were satisfied, and the* *disciples picked up twelve basketfuls of broken pieces that were left over.* [21] *The number of those who ate was about five thousand men, besides women and children.*

Food IS power and influence! When we moved to Panama to build a prototype commercial greenhouse farm, we used the power of food to feed the poor. Our first harvest of 2,500 pounds of bell peppers came in. We loaded them up in 30 lb. bags and drove them to town mayor's offices, children's orphanages, schools and feeding centers. Every bag of peppers that we gave away - gave us automatic influence with the leaders of those institutions, as well as with the mayors in their cities. I remember looking at many of their faces as they were in shock. Why did these gringos just bring us 100 pounds of peppers to us free of charge? And with no strings attached? And they are supporting and helping us in our jobs to try to help the community?

To the mayors and school principals we said, "We want to help feed the poor today with our farm harvest. But we don't know who the poor are. And we know that you are fully aware of the most needy in your schools and communities. So we want you to give these peppers out in your name. Please don't tell them that the peppers came from us. We are giving to you, and you can give to the people. We just want you to know that we support you and are thankful for all that you do for our community."

I'm not sure that some of these people had ever seen such a thing before, as it's not customary to give a lot of food away in our region. But we did it as an act of obedience to God. We felt that He led us to do this thing, not knowing fully why. We wanted nothing in return. But great influence came our way resulting in great favor in the entire region and nation. We supernaturally had a bridge built over our farm river, providing better access. Electric poles showed up from an unknown donor to bring electricity to our farm (which previously ran only on a generator). People came to help and volunteer their time. International doors continually open to us. Supernatural supply comes to our lives and ministry from every direction. This continual outpouring comes to us because we give to the poor. How does it happen? I don't know for sure. I just would never want to fall outside of this divine river of love, mercy, and supply.

When I was a young preacher in St. Louis before ever really giving food away in larger amounts, I had an opportunity to meet with a great international evangelist while in the Philippines. I had a unique opportunity to eat lunch with him while on a missions trip. This man had filled major stadiums around the world, he had seen major miracles in his ministry, and at that time he was feeding around 200,000 children per day worldwide. And little ole' me was going to have lunch with him one on one. I geared up for the best question I could possibly ask him. When we met for lunch, I shot my question to him. And his answer changed my life.

I said, "Sir, if there was one thing that you could attribute your great success to, what would that one thing be?" He replied, "I will tell you, but you are not going to expect the answer that I am going to tell you."

He went on to build the story line. He said, "You know, I'm not a very smart person. I am certainly not the greatest preacher. I get my Bible stories mixed up and might have Joshua building the ark and Moses crossing the Jordan. I'm really not a very nice person at the end of the day, and am actually pretty harsh and abrupt. If you and I worked together for very long, I'd probably offend you by something that I would say to you.

I'm actually pretty selfish and I don't know why God would have chosen me to work for Him. I'm a man of many sins and I struggle with that every day. But after all of these problems I have this one thing in my favor."

He then paused and I waited intently listening to every word he said.............

After the long pause, he then continued, "I feed widows and orphans. I help God fulfill His promise on this earth as He said, 'I will be a father to the fatherless and I will be a husband to the widow.' And I have found that as long as I help God get supplies to His wife and child, and help Him to fulfill His promise on the earth, then He in return, makes my dreams come true."

He continued, "I love to preach the gospel, pray for the sick, and travel around the world. I love to fill stadiums and see people come to Jesus. And He allows me to do all of this along with all of my mistakes. But I'm convinced it's only because I feed His widows and orphans, that he allows me to do the things that I love."

That answer wasn't what I was expecting, but it was the deepest, most impacting wisdom that I could have ever asked for. After that meeting, I made a personal decision that a portion of everything I did in my future would go towards widows and orphans. I would also strive to help God fulfill His promises in the earth. I would become a Global Food Provider to the poor, the widow and the orphan. I didn't know what that meant at first or how it would play out. But I knew that it started with multiplying our little church pantry in St. Louis, Missouri. And that word delivered into my heart "in season" has grown since then. I would have the guts to believe God for that first semi truckload of food to give to the poor, which turned into three truckloads (story above). And that faith would open doors that have never closed since that time.

Today, decades later, I can honestly say the same thing as this man told me.

You know, I'm not a very smart person. I am certainly not the greatest preacher. I get my Bible stories mixed up from time to time. I'm really a pretty harsh and abrupt person. I'm actually pretty selfish and I don't know why God would have chosen me to work for Him. But I do have this one thing in my favor. I feed widows and orphans. I help God fulfill His promise on this earth. And I have found that as long as I help Him get supplies to His widows and orphans, then He makes my dreams come true. I don't know how else to say it. Maybe:

Proverbs 19:17 (MSG) *[17] Mercy to the needy is a loan to GOD, and GOD pays back those loans in full.*

Or........

Proverbs 22:9 (MSG) *[9] Generous hands are blessed hands because they give bread to the poor.*

Or.......

James 1:27 (NIV) *[27] Religion that God our Father accepts as pure and faultless is this: to look after orphans and widows in their distress and to keep oneself from being polluted by the world.*
"Peter", Jesus said, *"Do you love me? Feed my sheep."*

The door to the supernatural awaits. May we be faithful to open the doors over us, our homes, communities, churches, businesses and nations. Especially with this cataclysmic food security crisis that's coming upon the world. I encourage you to embrace the day, seize the moment, hoist the sails, stretch out your faith and ask God for a bountiful supply from Heaven. Help Him fulfill His promises in the earth and you will become a Global Food Provider!

Join our "Agriculture College" online educational project, teaching leaders to start farming and gardening projects in their communities, enter the commercial food industry, and feeding the poor, downcast, widows and orphans.
www.globalfoodproviders.com .

Global Food Providers

Chapter 6: Agricultural Evangelism, An Open Door

Psalm 2:8-9 (GW) *[8] Ask me, and I will give you the nations as your inheritance and the ends of the earth as your own possession. [9] You will break them with an iron scepter. You will smash them to pieces like pottery."*

A door is now standing open to the nations most closed to the gospel, those most resistant, and those farthest and hardest to reach. The door is open through agricultural and business entrepreneurial education. In the darkest of regions where the gospel light has not been allowed access, we find people who are enslaved in their minds and given to mass poverty, both in mind and natural supplies. There seems to be a curse on their farm ground. Harsh dictatorships, Socialism and Communism rule over some of their governments. Many hindrances of the human soul have stopped peoples' ability to multiply, squashed creativity, and destroyed healthy business environments. Agriculture is also greatly hindered as people lose the will to work hard for little or no reward.

If we will take a good look at the hardest to reach nations we will find open doors everywhere. Agricultural and business education are desperately needed. Broken societies need fixed. From fresh water supplies to food chain provision. The people need job skill training, language education, and business classes. Anyone willing to go in and provide these opportunities has a wide open door into the hardest to reach nations that are closed to the old wineskin/bottle way of preaching the gospel.

Food is power. Whoever has the food has the power.

I call it Agricultural Evangelism. It involves taking the wisdom, knowledge, and insight to grow food and educate the people to provide for their own countrymen.

With food insecurity ramping up on a global level, doors are wide open for agricultural education and training to come into the most closed of nations. It would behoove us during this season to send a fresh wave of missionaries to Vocational Agriculture school, auto and farm equipment mechanics training, food packaging and processing training, truck driving school, and much more. All of these job skills are absolute open doors into nations with food insecurity. When formerly resistant people are hungry and their farms and businesses are wasting away, they could care less if you are Christian or not. If they see that you're willing to help them to conquer an agricultural, business, or societal problem, they will be grateful for your willingness to help. They will invite you in with open arms.

Deuteronomy 8:18 (KJV) [18] But thou shalt remember the LORD thy God: for it is he that giveth thee power to get wealth, that he may establish his covenant which he sware unto thy fathers, as it is this day.

The Lord has given us power to get wealth. Those two words are interchangeable. He gives us wealth to get power, as well as power to get wealth. And part of the power that He gives is wisdom, insight, business capacity, and the ability to see a problem with capacity to fix it. And the greatest wealth that a church planter can get is relational capacity with people.

Indeed a person with the ability to fix a farming problem is a very wealthy person in the agriculture world. The wisdom that this person can provide is invaluable as new ways are taught to grow food, new systems and ideas grow more food, and new secrets are shared from God's Kingdom to bring light for food production. Indeed God wants to use our church planters, missionaries and business people to establish His covenant in the earth.

Luke 10:2 (NIV) [2] *He told them, "The harvest is plentiful, but the workers are few. Ask the Lord of the harvest, therefore, to send out workers into his harvest field.*

We need to find "Sons of Isaachar" in our relational spheres of influence. These gifts from God are resident in men and women of business, agriculture, science, commerce and industry. They can easily read into times and seasons, and they know the best way forward for farms, entrepreneurial endeavors and businesses. They have a gift for seeing a problem and knowing exactly what to do in order to fix the problem. These are the people that we need to thrust into the harvest fields of the world. Their gifts will shine in dark places and will cause men to come to the light of the glorious gospel!

King David was surrounded by those gifted with these abilities. They were needed to help establish David's kingdom and throne. They knew times and seasons and what ought to be done for success.

1 Chronicles 12:32 (KJV) [32] And of the children of Issachar, which were men that had understanding of the times, to know what Israel ought to do; the heads of them were two hundred; and all their brethren were at their commandment.

This is a new day with new opportunities and open doors to reach into areas never accessible before. One major door available is agricultural evangelism. We must take the gospel along with the capacity to grow food and educate others how to do the same. We will build relationships and make disciples of those who receive the gospel.

I highly recommend that church planters of today consider taking with them, sons of Issachar, business and agricultural people, industrial professionals, and those with keen capacities to see societal problems with knowledge of how to fix those problems.

Church planters can literally go into closed nations and start a farm. A farm project will provide jobs and education to the community for food security. Food production will be purchased by the people. Relationships will be built with workers and their families, with business people and shop keepers who buy and sell your produce, and with neighbors who purchase your food. They will watch you and how you conduct your business with high principles, ethics and with care for the community. They will see you having concern for the poor and feeding them. They will see you offering a helping hand to those in need. Whether helping to repair a neighbors' roof, volunteer to help at the local school, or help paint and clean the local park. Your life and light will shine before these people. They will open up their hearts and lives to you. Food will become the power that brings you authority in your community. Mercy in operation towards the poor, the downcast, and the needy will cause you to stand out among the people as an exceptional people who must have been sent by God. And the power of the miraculous will follow you as you love and care for the people you have planted your crops among. When you pray, God will answer. And the people will learn to trust your God, and will desire to follow Him, as you do.

Not only will the farm be your entry into the community, it will also be an avenue of financial provision for your family. Farm produce sales will bring you cash flow and will help you pay your bills. As you become a permanent installation in your community through agriculture, your influence and reach will be far and wide. And if you bring new farming concepts and innovations to the community, they will ask you to come help them from farm to farm, community to community, and from seminar to conference. God will fulfill His promise to you as he promised Abraham:

Genesis 12:2 (KJV) [2] And I will make of thee a great nation, and I will bless thee, and make thy name great; and thou shalt be a blessing:

I can testify personally that our farm, started in Chiriqui, Panama, has made major inroads to the entire farming community and province. We have made friends with hundreds of farmers, producers, product specialists and produce buyers. We have a very special relationship with mayors, school principals, teachers, college professors, government regulators and office personnel, bankers, restaurant owners/managers and their employees, auto mechanics, repair specialists, and people everywhere that our farm operation connects with daily, weekly and monthly. Our neighbors are our friends. Our employees are family. We will fight to protect one another and we continually help one another like family and true friends should. We spend time together and extend the love of God wherever possible. Our light has reached farther through our farm than any traditional Sunday morning church building could ever reach.

When we started our prototype farm project, the Lord revealed clearly to us that we were going to do things backwards from the old normal way of church planting. We would start the farm first, and then eventually we would start a Bible college to train Christian leaders. Then we would start a church with the leaders that had been raised up through the Bible college.

The farm is a model concept for the hardest and most closed nations on earth. The kingdom is inside of you, and needs to be inserted into their society. But it needs to come in through agriculture, business or education, providing powerful answers to their worst problems. When you provide the answers along with food provision for their markets and towards their poor, your influence will break down the walls of resistance and you will win great favor among men. Your mustard seed will grow to the greatest of trees for the community to take refuge in. And your kingdom leaven will work into the hearts of the people, and the community will be converted by the power of God unto salvation.

Matthew 13:31-33 (NIV) [31] *He told them another parable: "The kingdom of heaven is like a mustard seed, which a man took and planted in his field.* [32] *Though it is the smallest of all your seeds, yet when it grows, it is the largest of garden plants and becomes a tree, so that the birds of the air come and perch in its branches."* [33] *He told them still another parable: "The kingdom of heaven is like yeast that a woman took and mixed into a large amount of flour until it worked all through the dough."*

Undoubtedly business, education, language and job skill training are open doors into closed societies. But because of the global food in-security that's reaching deep into these closed nations, it's clear that the Lord is putting wind in the sails of anyone who wants to step in through agricultural evangelism. Indeed the door is open wide.

As a side note, we are working hard to provide inspiration, agricultural training, relational capacity and farming assistance for those who are willing to take the supreme challenge. Join us in our quest to build up, train, and send agricultural evangelists to the hardest and most closed nations of the world. www.globalfoodproviders.com

One of my favorite songs from Steven Curtis Chapman tells the story. The words to "The Great Adventure" are the words we live every day as agricultural evangelists.

"Saddle up your horses. We've got a trail to blaze. Through the wild blue yonder of God's amazing grace. Let's follow our leader into the glorious unknown. This is the life like no other. This is the great adventure."

Chapter 7: PHARAOH HAS A PLAN

In the days of Joseph, his successful interpretation of Pharaoh's dream put him at Pharaoh's right hand. He was responsible for gathering food during the seven fat years and preparing for the seven lean years to come. He would be faithful to help people when the seven lean years of famine hit Egypt. But his boss would concoct a devious plan to capitalize on the famine and eventually own everyone's land in exchange for food to keep them alive. Pharaoh was a business man and would use the famine to gain great wealth. His plan was diabolically wise and calculated how long people's savings accounts would last them before being forced to sign over their farms, lands and homes. In exchange for being cared for through the end of seven years, Pharaoh would trade short term provision for long term land ownership. Why not? He knew the future. But he used it for his own increase rather than to be an instrument of mercy towards the "soon to be" poor.

The spirit of Pharaoh has a diabolical plan for mankind today which mirrors the Old Testament Egypt story. There's an old saying that a government leader spoke not too long ago which said, “We don’t want to ever let a crisis go to waste.” In other words, let us seize upon a humanitarian crisis and make it an opportunity to increase our authority, power, and capacity.

Pharaoh is ramping up to seize control of people, their farms, homes and lands, and entire nations as this new global food insecurity ramps up. We see an attempt of the spirit of Pharaoh to tax our carbon footprint and energy use. Governments are looking for ways to tax our water and air, to introduce regulations that limit and control how we live and what we eat. In the name of climate change, new ideas of taxation and control are being introduced around the world. This is simply Pharaoh at work to seize on a crisis and turn it into an opportunity.

You can contact Dr. Daniel Daves on Facebook, Instagram, Twitter, or by email at: info@globalfoodproviders.com.

Chapter 8: FINAL THOUGHTS

This is a time for us to walk in all gravity, sober and vigilance, and to provide Heaven's answers to needy people without taxation and seizure of wealth. While we are encouraged to help the people as Joseph did, I believe that this time around, we need to provide answers that will stop the crisis, the famine, and the seizure of lands by the spirit of Pharaoh.

Through wisdom, knowledge, personal sacrifice and great faith we can: lead the people to God's glorious kingdom, bring them to His supply and provision, and bring the light of the glorious gospel of Jesus Christ into lives, homes, industries, governments and societies. There is a higher way, and we are the ones to provide that way.

Isaiah 62:10 (GW) *10 Go through! Go through the gates! Prepare a way for the people! Build up! Build up the highway! Clear away the stones! Raise a flag for the people!*

And……

Matthew 28:1-10 (NIV) *1 After the Sabbath, at dawn on the first day of the week, Mary*

Magdalene and the other Mary went to look at the tomb. [2] There was a violent earthquake, for an angel of the Lord came down from heaven and, going to the tomb, rolled back the stone and sat on it. [3] His appearance was like lightning, and his clothes were white as snow. [4] The guards were so afraid of him that they shook and became like dead men. [5] The angel said to the women, "Do not be afraid, for I know that you are looking for Jesus, who was crucified. [6] He is not here; he has risen, just as he said. Come and see the place where he lay. [7] Then go quickly and tell his disciples: 'He has risen from the dead and is going ahead of you into Galilee. There you will see him.' Now I have told you." [8] So the women hurried away from the tomb, afraid yet filled with joy, and ran to tell his disciples. [9] Suddenly Jesus met them. "Greetings," he said. They came to him, clasped his feet and worshiped him. [10] Then Jesus said to them, "Do not be afraid. Go and tell my brothers to go to Galilee; there they will see me."

May the Lord bless our journey into the great adventure. And may we preach the gospel to all nations, tribes, people, and tongues so the whole world may know the saving grace of our Lord and Savior Jesus Christ.

Remember, “Food is power. And whoever has the food has the power.” – Dr. Don Matison

\- End

Dr. Don Matison & Dr. Daniel Daves together at one of their Guatamala children’s feeding projects

World Missions Trip

Come join us on an agricultural vacation or bring a missions team to help us build, feed children, and preach the gospel. Join us on an international trip to train church planters. Find out more information at www.globalfoodproviders.com

Dr. Daniel & Tracy Daves

Global Food Providers

Working with the precious indigenous people of Central America, feeding their poor and training them with job, business, and language skills.

Providing food, farming supplies, seed and education.

Distributing fresh food to widows and orphans.

The joys of feeding the innocent and hungry.

Church, community, and evangelistic services towards very spiritually hungry people.

Learn farm techniques for the mission field.

Work with schools and orphanages who are generally wide open to the good news and mercy feeding operations.

Memories: Photos of the first “3 semi truck” evangelistic campaign in Fairgrounds Park, St. Louis, MO. USA. as mentioned in chapter 5.

Unloading fresh and frozen food as fast as possible.

Coordinator Larry Key in front of food bank truck.

Refrigerated Solution

Used Clothing Give Away.

1,000 hotdogs & hamburgers.

Children's outreach

Successfully unloaded truck #3. Sore muscles!

Our dear friend Bishop Arthur Kelley in the background, who set up all of the church connections.

40,000 pounds of fresh food per truckload x 3 trucks!

Thirty five churches working together in unity. Red, yellow, black, and white all standing against crime, community fear and poverty. The community gathered and received a divine outpouring of love and provision. The response was overwhelming. There was so much food at the end of the event, that we stacked church pantries full!

Matthew 15:36-37 (MSG) *[36] He took the seven loaves and the fish. After giving thanks, he divided it up and gave it to the people. [37] Everyone ate. They had all they wanted. It took seven large baskets to collect the leftovers.*

From The Philippines Story of Miracle Food Provision in chapter 5.

One of the Philippines feeding projects where the food miraculously multiplied in front of our eyes!

See Dr. Daniel Daves' training materials for business, ministry and finance.
www.doctordanieldaves.com

Made in the USA
Monee, IL
04 November 2026

67377618R00036